The Energy Cycle

RETHINKING GRAVITY

"For decades Einstein attempted to develop a unified field theory, a model of the universe that would explain gravity and electromagnetism as manifestations of a single force, connecting the movement of planets and stars with the operations of the tiniest subatomic particles." (Lacayo, 2014, 9)

Martin O. Cook

Edited by Stephen Gibbons

Insightful Contributions by David Miller, Monte Mower, Jeffery Davis, **Wendy Cook**, Stephen Gibbons, Jack Hylton Jr., Bryan McPherson, Randy Dean, Mike Scott, John Cook Sr., Sally Cook, Kylene Schramm, Thomas Cook, Sarah Cook, Rebekah Cook, Isaiah Cook, Kathryn Cook, and Kelsey Dean Grace Cook

Copyright © 2014 Martin O. Cook
All rights reserved.
ISBN-13: 978-1500723101
ISBN-10: 150072310X

The universe operates through cycles. A cycle is a pattern that repeats itself over and over again like the water cycle. The water cycle keeps the flow of water moving in a continuous pattern from the oceans to the mountains and back to the oceans. An energy cycle is a repeating pattern that keeps the flow of energy moving through a galaxy. The energy cycle, like the water cycle, can be broken down into steps in order to easily explain the overall picture of how it works. It is intuitively easy to understand.

Step 1: Energy continuously flows out from the center of a galaxy. Stars already in motion absorb this energy. This energy is the gravity that keeps a star's motion in an orbital pattern around the center of the galaxy.

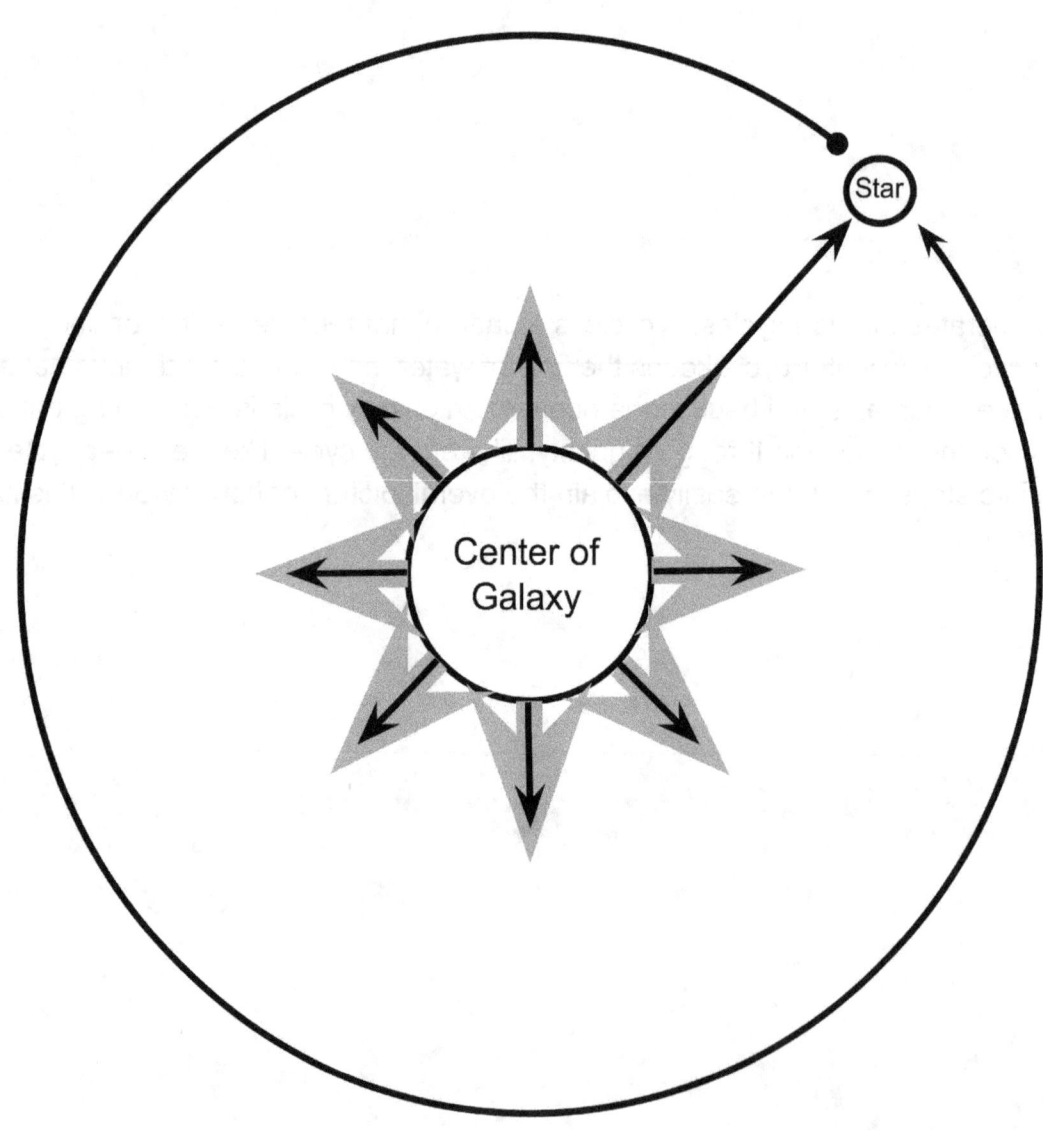

Step 2: As a star continuously absorbs energy, it also continuously emits energy. It does so proportionately through its spherical surface. This causes planets and other objects already in motion to orbit around its spherical shaped body as it orbits around the center of the galaxy.

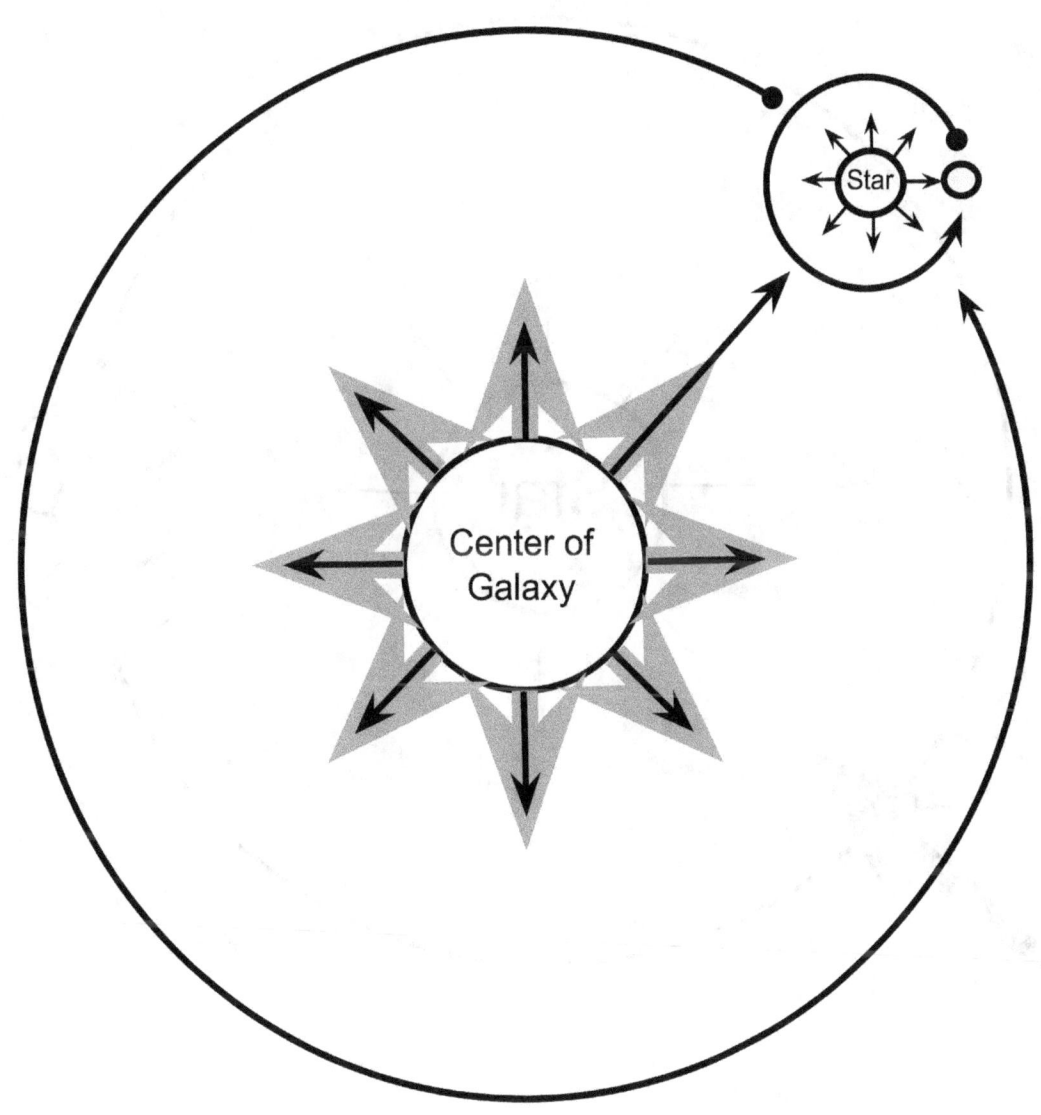

Step 3: Spherical bodies, like planets, continuously absorb energy coming from the star they orbit. This energy is part of the gravity process that keeps them orbiting the star. As planets absorb energy, they also emit a steady flow of energy through their spherical surfaces. This causes other objects like a moon (that is already in motion) to orbit the planet. The emitted energy also keeps objects like balls, cars, bikes, humans, and animals that are already in motion from floating off into space.

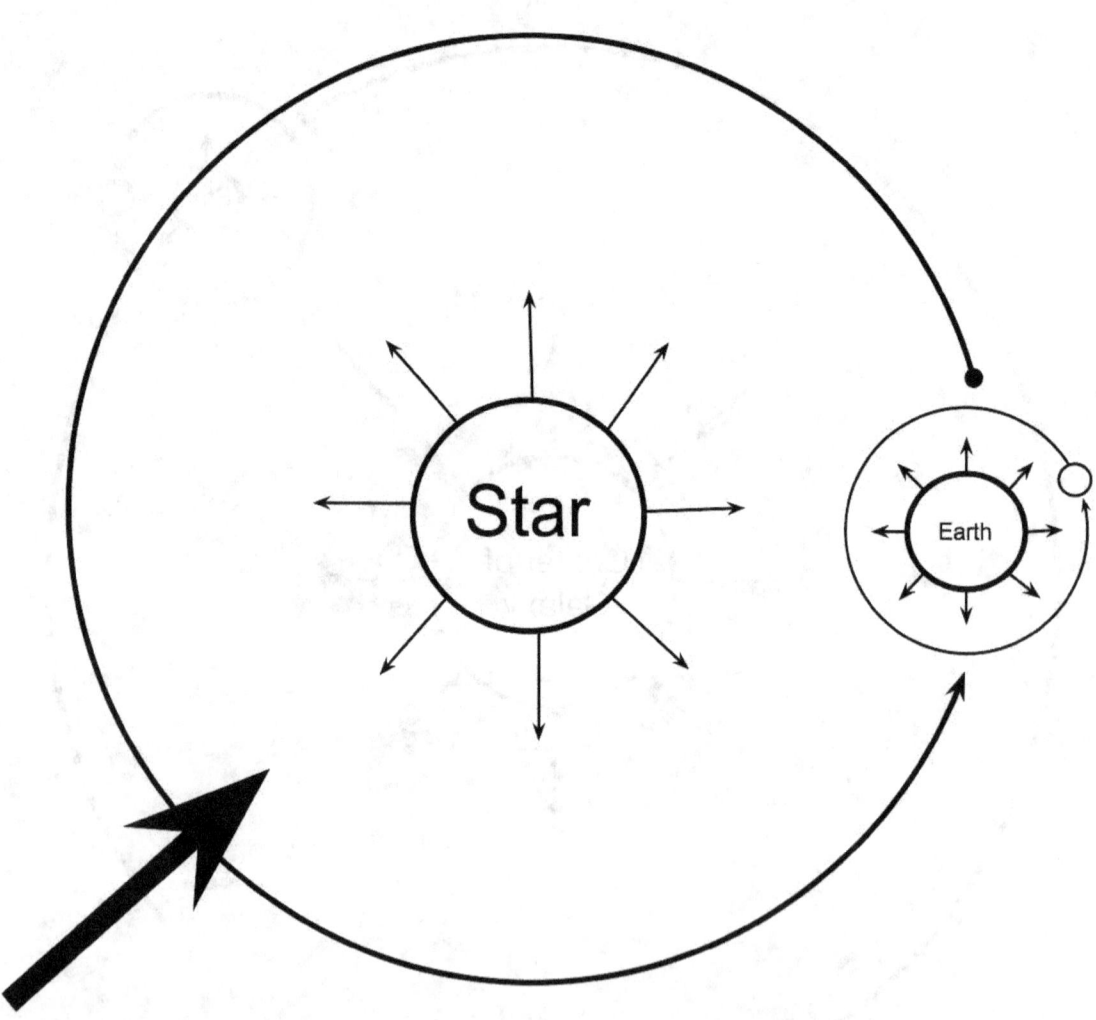

Step 4: Much of the energy that is emitted by the center of the galaxy makes it back to the center of the galaxy, where it is absorbed and redirected back out into the galaxy.

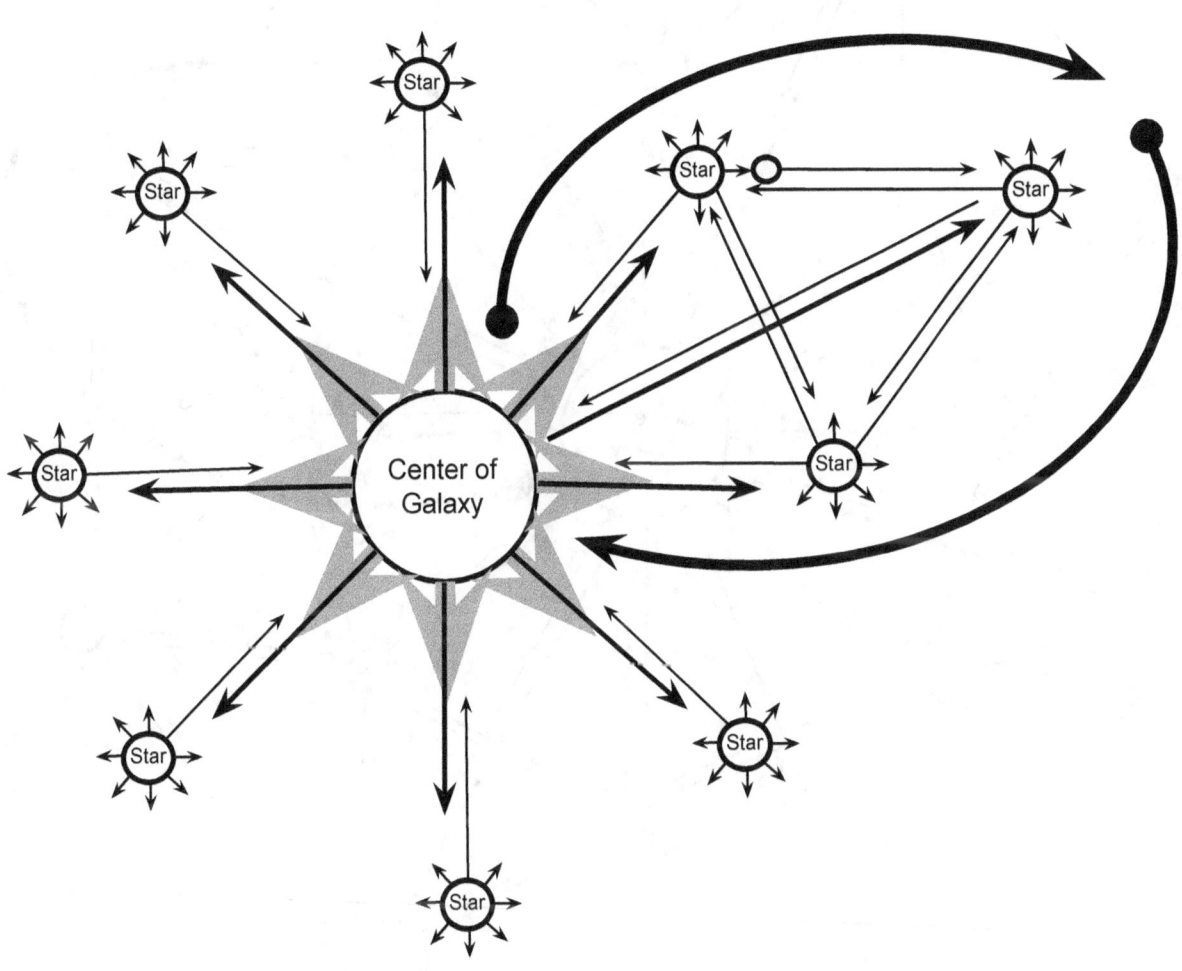

A complex web of spherical bodies sustains the energy cycle within each galaxy.

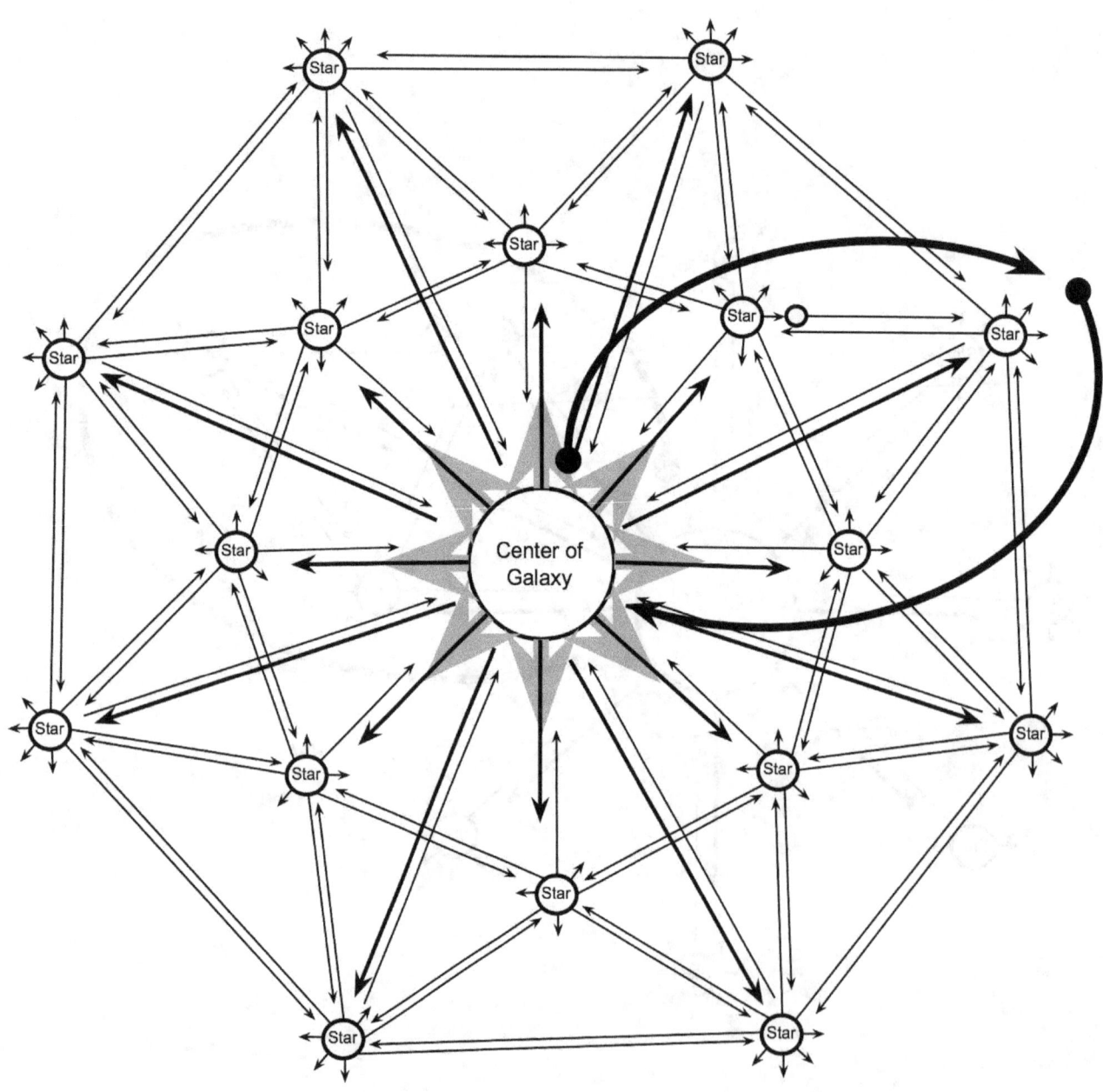

Step 5: Energy that escapes the energy cycle of a galaxy feeds into other galaxies. In turn, galaxies receive energy from other galaxies to replace the energy they have lost. Our galaxy receives lost energy from other galaxies to replace its lost energy. Galaxies, like stars, play an important role in keeping the balance of energy flowing throughout the universe.

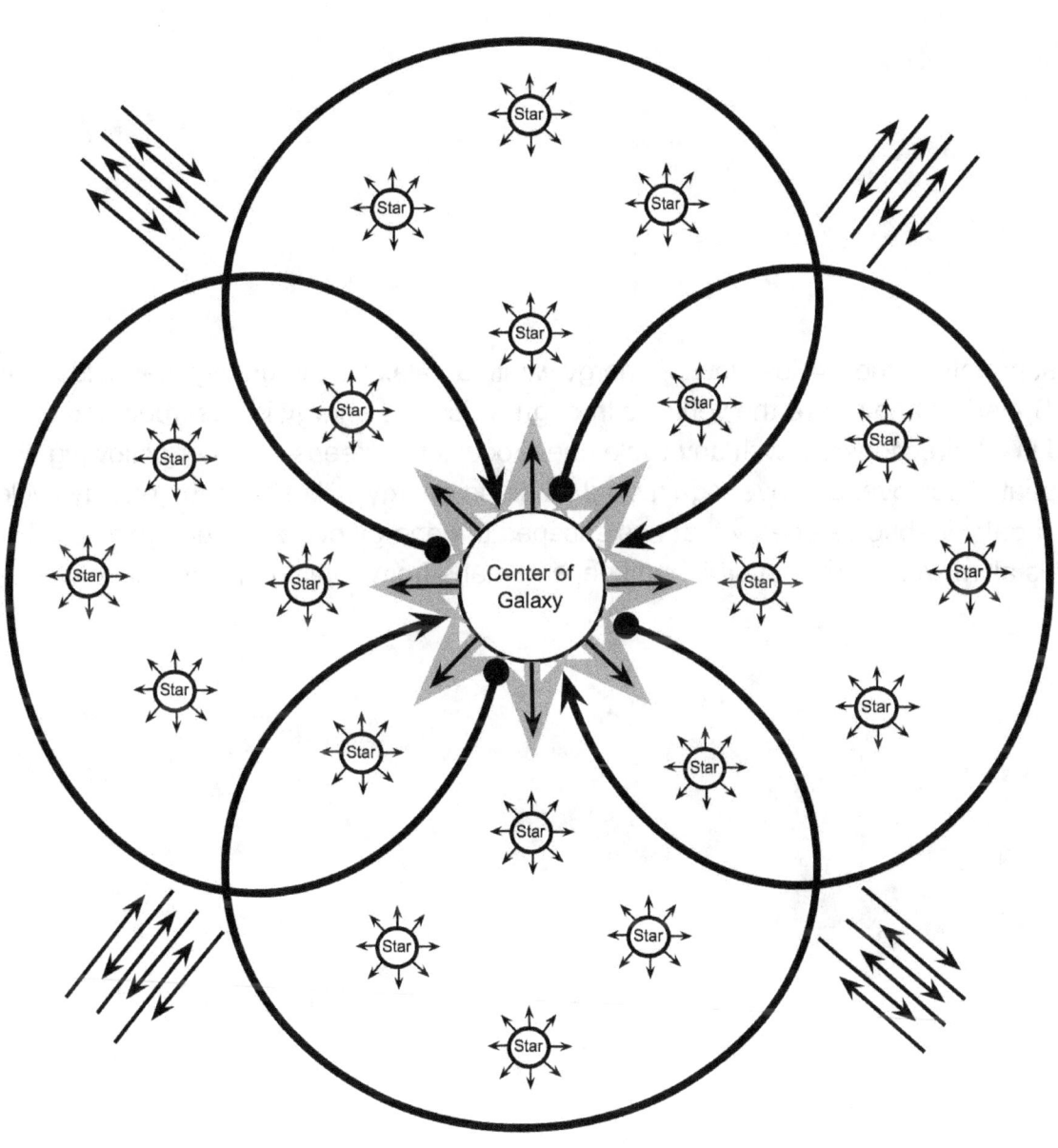

In summation, the cyclical flow of energy within a galaxy is the gravity that holds a galaxy together. Spherical bodies are the vehicles through which this energy is absorbed and then redirected in all directions. As with any cycle, the process that keeps the energy flowing within a galaxy repeats itself over and over again. And as some energy will escape the energy cycle of our galaxy, our galaxy absorbs energy that has escaped the energy cycle of other galaxies. The Energy Cycle holds all moving bodies in orbit and keeps humans from floating off into space.

Momentum, relativity, and gravity are functions of the energy cycle as evaporation and rain are functions of the water cycle. Momentum, relativity, and gravity as functions of the energy cycle are part of the same science of motion that hold planets in orbit and prevents people and things from floating off into space. There are two essential principles at the heart of the energy cycle.

1. The role of the atom in momentum, relativity, and gravity. (**Part I**)
2. How spherical bodies absorb and redirect energy within a galaxy. (**Part II**)

Part I

The Role of the Atom in Momentum, Relativity, and Gravity

The Illusion of Rest

Sir Isaac Newton declared that an object in motion tends to stay in motion, while an object at rest tends to stay at rest. What does it mean that an object is at rest?

Think of two astronauts experiencing a continual free fall in the international space station as it orbits the earth. As the astronauts appear to be floating within the confines of the space station, one astronaut passes an orange to the other astronaut. As the orange makes the trek through space, its motion is visible to both astronauts. When the astronaut receiving it places it in space next to him, the orange now appears to have stopped moving as it takes a position of rest next to the astronaut who placed it there. Is the orange really at rest?

What appears to be at rest to the astronaut who just placed it in space next to him is really freefalling around the earth at about 17,500 miles per hour. The orange that appears to be at rest is still in motion through space.

Although objects can appear to be at rest, it doesn't mean they are not moving through space. They still have momentum. Newton's law of motion that states that *an object in motion tends to stay in motion* applies to all objects, even objects that appear to be at rest, such as an object next to an astronaut in the space station or objects that appear to be at rest on the earth's surface.

To emphasize this point, imagine a switch that turned off the gravity that accelerates objects into the earth's surface. With the effects of gravity shut off, the slightest nudge to any of these objects would cause these objects to move away from the earth's surface. The nudge wouldn't begin the motion; it would only initiate a slight variance of the motion they already exhibited when they were moving through space attached to the hip of the earth by the effects of gravity. As these nudged objects, once bound to the earth's surface, continued in their new momentum path, moving away from the earth's surface, one would witness that these objects always had motion. Their motion was cloaked by their gravity-bound status to the earth's surface. Through the effects of gravity, their motion became synchronized with the motion of the earth.

All masses are always in motion, even masses that appear to be at rest. The "illusion of rest" is the greatest stumbling block to a viable explanation of how gravity really works from an atomic perspective.

The Atom and Motion: (*Each atom is a self-regulating confinement of energy.*)

How does an object in motion stay in motion? We know that a car in motion eventually runs out of gas and stops. Why doesn't an asteroid orbiting the sun run out of gas? It just keeps going and going and going. What fuels the inertia of mass? In other words, how does an object in motion stay in motion? Does mass, such as an asteroid orbiting the sun, just mysteriously stay in motion through space without any explanation other than external forces that can speed it up or slow it down? No, it is on the atomic level that mass, which is made up of atoms, gets its motion.

Atoms don't just mysteriously move through space. The energy making up an atom fuels its momentum through space, even when it appears to be at rest in a mass on the earth's surface. Any change in an atom's momentum is accompanied by an energy change to compensate exactly for the momentum change.

The key here is that the confined energy making up an atom also accounts for its momentum, so any change in momentum must be accompanied by a change in the energy driving the momentum.

As a famous saying goes, this is where the rubber meets the road. This single point, that **each atom is a self-regulating confinement of energy that regulates changes in momentum though energy adjustments,** forces us to rethink the role of the atom in momentum, relativity, and gravity.

The Atomic Model of Motion

The Atomic Model of Motion explains the role of the atom in the momentum, relativity, and gravity of mass. The atomic model of motion also sets the foundation for understanding how energy is recycled through spherical bodies.

The Atomic Model of Motion Overview

1. <u>Quantum Momentum</u> (or the role of the atom in the momentum of mass) is the equilibrium state of momentum where energy is neither absorbed nor emitted. This is the reason that an object in motion stays in motion.

2. <u>Quantum Adjustments</u> (or energy adjustments) are changes in momentum (acceleration and deceleration) that are accompanied by the absorption or emission of energy to exactly compensate for those changes. The science of motion is as exact as the science of chemistry. All changes in momentum can be explained by the addition or subtraction of energy.

3. <u>Quantum Relativity</u> (or how the energy of atoms and masses change as they go from one inertial frame to another) explains Galilean relativity from an atomic perspective. The sustaining energy driving all inertial frames is measurable and accountable on an atomic level using an understanding of quantum momentum and quantum adjustments.

4. <u>Quantum Gravity</u> (or the role of the atom in gravity): The acceleration of mass we call gravity is nothing more than the absorption of energy into the atoms of mass that accelerates already moving mass in the direction of absorption. The result of absorbed energy is acceleration. The effects of increased energy within mass that causes the acceleration are easily observable.

1. Quantum Momentum: The Role of the Atom in the Momentum of Mass

Quantum momentum is the reason why an object in motion stays in motion.

The momentum of each individual atom drives the momentum of any given mass.

The atom is the never-ending energy that keeps all mass in never-ending motion. This is how planets continuously orbit the sun without having to be rewound or refueled. Even objects appearing to take a break on earth's surface still cruise at nineteen miles per second as gravity yields them earthbound. Turn off gravity and these objects are no longer attached to the earth's surface. Give them a little nudge and they will float off in uniform bliss until an interaction with some other form of energy redirects their route or changes their speed.

Quantum Momentum is the equilibrium state or uniform motion of any atom and is the foundation for all motion from the simplest atom to the greatest of masses. The motion and momentum of any given atom is independent of the motion and momentum of all other atoms. It is the bonded relationship of the independent motion and momentum of the atoms making up a mass that gives that mass its motion and momentum through space.

From this point forward, I will periodically use the theme of yoked horses to explain the role of the atom in momentum, relativity, and gravity.

The best way to describe the Quantum Momentum of an atom is to take an adventure into the nucleus of an atom. Visualize a proton and a neutron yoked together like two horses yoked together pulling a carriage. Visualize the movement of the atom as the movement of these two horses. The yoke acts like the strong force that synchronizes their independent movements. With the snap of a whip or a pull on the reigns, the horses speed up, slow down, or change directions. Now imagine that these horses never stop moving. When forces act upon them, they speed up, slow down, and or change direction, but like the horses pulling Zeus' chariot, they never run out of energy.

In the emergence of an atom, energies form the atom's structure (yoking the horses together) while simultaneously engaging the atom's momentum through space, (the synchronized movement of the two horses when yoked). The two, structure and momentum, are inseparable.

As atoms form bonds with other atoms to create masses, the motion of each atom is linked together with the motion of other atoms to create the motion of the mass. The atoms that make up any mass share an order of movement that orchestrates that mass's movement through space.

A good way to visualize the **quantum momentum** of mass is to think about objects in momentum in the space station while it is orbiting the earth. Because the space station is in a free-fall, the independent momentum of all objects can clearly be observed. When an astronaut pushes an object, such as a bag of water, he is pushing all the atoms that make up that bag of water. When the force ceases, the atoms settle into the equilibrium of uniform motion. The atoms of the bag of water continue in the same direction until they bump into a wall of the station. As they run into the wall, a chain reaction transpires (which will be discussed in detail in the next section) until each atom experiences a momentum shift that leads to a new speed and direction. This new path continues until a force again acts upon all the atoms making up the bag of water. These atoms continue their new momentum uninterrupted even if the space station were to suddenly vanish from around them.

Any mass that appears to be at rest is just sharing the same motion with the object it appears to be resting on. Objects that the astronauts place in the air next to them—such as a toothbrush—stay in the same place. This is because the atoms making up the toothbrush have the same momentum through space as the atoms making up the astronaut. Their momentums are synchronized, moving at the same speed and in the same direction. Another example of the shared uniform motion of separate masses not connected to each other is objects within a car but not connected to it. As the vehicle accelerates, all the objects within the car go through internal changes to accommodate for the acceleration of the vehicle they appear to be resting in. At each new speed, all the objects in the car obtain the same uniform or quantum momentum as the car. When the car crashes into something, as its momentum is instantly changed, each object that appeared to be resting in the car continues in their own independent momentum until they crash into something such as the back of the seat, the dashboard, or the windshield. Each object has its own unique inertial momentum independent of all the other objects, including the car itself. Hence the importance of seat belts that physically attaches our momentum with the momentum of the vehicle in which we are riding.

What makes the quantum momentum of atoms and mass difficult to grasp is we continually see objects at rest on the earth's surface with no apparent motion whatsoever. We see rocks that have settled to the ground, furniture resting in houses, or a computer sitting on a desk. All of these seem to have no motion or momentum. The computer I am typing on does not appear to have any momentum so how could the atoms that make it up have momentum? This appearance of objects at rest is the same illusion that fooled Galileo and Newton and is still fooling physicists to this present day. The computer I am typing on is in a continuous state of acceleration towards the earth and only appears motionless because it is temporarily experiencing something similar to terminal velocity. Due to the effects of gravity, the uniform motion of the computer's atoms is riding on the uniform motion of the atoms of the earth. If the effects of gravity could temporarily be shut off, the momentum of the atoms of the computer through space would be more apparent. The atoms of the computer would continue to move through space at the same momentum of the atoms of the earth without being connected to the atoms of the earth because they would no longer be accelerating into them. Now if you gave the computer a push, it would slowly drift away from the earth, accentuating its own momentum through space, the synchronized momentum of the atoms from which it is made.

Quantum momentum is the inertial momentum of mass through space, which is caused by the synchronized momentum of the atoms from which mass is made. The speed and direction of mass through space remains unchanged just as Newton observed until an unbalanced force or energy acts upon the mass. Then the speed and direction are altered, but never at any time do the atoms of that mass stop or rest.

So what is the difference between an object at rest and an object in motion? Nothing. A body at rest shares the same motion as the body it appears to be resting on but their momentums through space are independent of each other. Just shut off the effects of gravity to prove this point.

2. Quantum Adjustments: Think of this section as a brief intermission that will be a useful tool to better understand quantum relativity and quantum gravity.

Quantum Adjustments are about the uneven forces that accelerate or decelerate mass. When mass accelerates or decelerates, quantum adjustments transpire within each atom to compensate exactly for these changes.

The momentum of mass does not change—a brilliant observation by Newton— unless acted upon by a force. The result of that force is a change in momentum. As changes in momentum occur, energy adjustments also transpire. Like the science of chemistry, the science of motion is exact. We just haven't figured it out yet like we have for chemistry.

As an atom's momentum increases, the atom will adjust its energy to exactly correlate with that atom's new momentum, and if an atom's momentum decreases, the atom will adjust its energy to exactly correlate with that atom's new momentum. These adjustments reestablish equilibrium within the atom as it goes through momentum changes. When the atom is in a state of equilibrium, that atom is in uniform motion and will maintain that motion until acted upon by another force or energy.

During momentum changes, each atom independently absorbs or emits the exact amount of energy to compensate for the momentum change. This regulates the momentum of an atom through space, whether the atom is by itself or is part of a collection of atoms in the form of a mass. All atoms experience quantum adjustments in energy in direct response to changes in momentum. This correlates perfectly with the conservation of mass-energy.

Quantum adjustments can better be explained by viewing each atom with a corresponding momentum pattern and energy level. The momentum pattern and energy level of an atom are two sides of the same coin that help explain changes to the continuous momentum of atoms through space. During the formation of an atom, protons, neutrons, and electrons unite in a synchronized dance to form and maintain the structure and motion of that atom through space. This structure, the atom's *momentum pattern*, describes the atom's perpetual momentum through space. The *momentum pattern* of an atom remains constant unless unbalanced forces act upon it. For this reason, I refer to an atom's uniform motion through space as a *momentum pattern* because this pattern continues until it is disrupted. As an atom's momentum changes, energy is proportionately absorbed, emitted, or adjusted to compensate exactly for the momentum change. Quantum adjustments are necessary to exactly compensate for disruptions to an atom's *momentum pattern*.

A good way to describe a *momentum pattern* is to go back to the visualization of horses yoked together. Think of one proton and one neutron as a pair of horses yoked together side-by-side walking at the same speed and in the same direction. Their combined movement is their momentum pattern. If we added additional pairs of horses to the first pair, so that each pair was lined up like a team of dogs pulling a sled, the combined movement of each pair of horses would be the momentum pattern of the combined team. Notice that no matter how many sets of horses we yoke together, each individual set of horses move at the same speed and in the same direction as all the other sets of horses, but with each additional set of horses added, the overall pulling power increases. Thus, the momentum pattern describes the speed and direction of the yoked horses, notwithstanding the number of sets.

Like the yoked horses, the nucleus of an atom could be described as yoked energies that simultaneously make up an atom's structure (the yoked horses) and momentum (the speed at which they are moving) through space. This energy, the energy that makes up the protons and neutrons of the nucleus of an atom, accounts for the motion of that atom through space. Like electrons, which can absorb and emit energy, protons and neutrons absorb and emit energy that regulates their momentum through space. As the linear speed of an atom increases, energy is simultaneously absorbed into every proton and neutron of that atom. As the linear speed of an atom decreases, energy is simultaneously emitted from every proton and neutron of that atom. (The electrons play a vital role in keeping the balance of energy within an atom by absorbing and emitting energy into and out of the nucleus and also into and out of the atom.)

The speed of any atom through space is determined by the *energy level* of each proton and neutron that makes up its structure or *momentum pattern*. As a force increases the speed of an atom, energy is absorbed into each proton and neutron. This increases the *energy level* of each proton and neutron of that atom. The same atom will now have an overall higher *energy level* per proton and neutron. Conversely, as a force decreases the speed of an atom, energy is emitted from each proton and neutron. This decreases the energy level of each proton and neutron. The same atom will now have an overall lower *energy level* per proton and neutron. The *energy level* of an atom (the speed the horses are moving) is in direct proportion to the energy level of each proton and neutron of an atom.

Energy level could be used in two different ways. There is the overall energy level of an atom (the total horse power) and the *energy level* of each pair of a proton and a neutron of an atom (the actual speed each individual horse is moving, which is also the same for all the horses). There is an important distinction. Let us compare an atom with one proton and neutron pair (one team of horses) to an atom with five pairs of protons and neutrons (or five teams of horses). If these two different atomic size atoms have the same momentum through space, then the *energy level* of each pair of protons and neutrons is equal (all the horses are moving at the same speed), but the overall energy level or amount of energy (total horse power) will be greater for the atom with five pairs than the atom with one pair. It will have about five times the overall energy, even though the atoms move through space at the same speed. The atom with the five pairs will have more inertia or resistance to change than the atom with one pair. Throughout the rest of this book, when I refer to the *energy level* of an atom, I am referring to the energy level of one pair of a proton and neutron. This is the *energy level* that describes its linear speed through space, notwithstanding its atomic number.

The protons and neutrons ability to absorb and emit energy allows for an atom to adjust its momentum when a force acts upon it. This holds true for each atom, notwithstanding the number of protons and neutrons in its nucleus. As the atom increases in speed or changes direction, energy is absorbed equally into every proton and neutron of the nucleus of that atom (all the horses start running faster). As the atom decreases in speed, energy is emitted equally from every proton and neutron of the nucleus of that atom (all the horses slow down). This is how the energy level (speed of one horse) and momentum pattern (the equilibrium state or steady speed of the entire team) are inseparably connected. A change in the momentum pattern of an atom (a force speeds it up or slows it down) initiates a simultaneous change in the atom's energy level (the speed at which each horse moves).

The horse example shows how the momentum pattern and energy level are actually a single process. As the *energy level* of each horse increases, the *momentum pattern* of the entire team increases by the same amount, and vice-versa. Imagine you have a team of six horses, with two horses yoked side by side, with another set yoked behind them, and the last set yoked behind them. The speed that each horse moves is the *momentum pattern* for that whole team. They all move at the same speed. When you apply a force, such as a whip to speed them up, or a pull on the reins to slow them down, you simultaneously change the energy level (speed) of each horse and the momentum pattern (speed) of the whole team. The *energy level* of one horse is the same as the

19

momentum pattern for the entire team. The energy level and momentum pattern for one set of a proton and a neutron is the same energy level and momentum pattern for all the other sets of a proton and a neutron within the same atom. When you change the momentum pattern, the energy level simultaneously changes. And vice versa, when you change the energy level, the momentum pattern simultaneously changes. They are two sides to the same coin.

Mass is a collection of synchronized atoms moving through space as the momentum pattern of each atom moves in concert with all the other atoms of that mass. Mass maintains a constant speed and direction until there is a disruption to the momentum pattern of each atom that makes up that mass. When the momentum patterns of the atoms of a mass are disrupted, energy is absorbed or emitted from the mass, from each individual proton and neutron of each atom. This reestablishes equilibrium within each atom within the mass to accommodate for the new speed and direction of that mass through space.

Visualize two masses colliding. The *momentum patterns* of the atoms of each mass are disrupted. As the *momentum patterns* are altered, the accompanying *energy levels* of the protons and neutrons of the atoms of each mass simultaneously adjust to accommodate the new *momentum patterns*. An absorption or emission of energy to exactly coincide with the degree that the *momentum patterns* were altered transpires, (each horse slows down or speeds up). As equilibrium is restored, the masses continue in their new momentums until a force or energy acts upon them again.

In summary, the momentum pattern of an atom and energy level of each proton and neutron that makes up the nucleus of an atom regulates the speed and direction of that atom through space. When momentum changes occur, quantum adjustments transpire to reestablish the equilibrium of quantum momentum. Since mass is the synchronized movement of individual bonded atoms, a momentum change in mass is a momentum change for every atom that makes up that mass. Every atom experiences a quantum adjustment. As there is no *something from nothing* in the cause and effect science of physics, changes in the momentum of mass must be accounted for by corresponding energy changes.

3. Quantum Relativity: How Atoms and Masses Change as They Go from One Inertial Frame to Another

Galileo learned that if a person were in a closed cabin of a ship moving at a uniform speed, he wouldn't be able to tell his relative motion to land by an appeal to the laws of physics. If he dropped an object, it would fall straight down. If he threw it up in the air, it would go straight up and straight down. In all, there was nothing he could do to determine his relative speed to land by applying what he knew about the laws of physics.

On the surface, the laws of physics appear to be the same for all inertial frames, but a deeper look into the microphysics of relativity will reveal differences on an atomic/quantum level for the same mass in different inertial frames. Since Einstein liked to use trains to explain special relativity, I will use a train example to explain quantum relativity.

If person A, traveling on a train holds a beanbag four feet above the floor and drops it to the floor of the cabin wherein he is riding, he will measure the beanbag to have fallen four feet. If person A decides to drop the same beanbag out of the moving train's window four feet above the ground, he will see the beanbag fall four feet straight to the ground in the same manner that he saw the beanbag fall four feet straight to the floor in the cabin of the train. Person B, standing on the stationary earth relative to the moving train, will measure the distance of the falling beanbag to fall more than the four feet as observed by person A. For person B, the beanbag will not only fall four feet to the ground as measured by person A, it will also fall at an angle proportional to the speed of the train. For this reason, each observer will literally measure the beanbag falling different differences. (See illustration—1). This is classic Galilean relativity, but it brings up an interesting dilemma: How can the same falling object literally travel multiple different distances such as this falling beanbag dropped by person A and observed by person B? The answer to this question will be found in the *momentum pattern* and *energy level* of each atom of person A, person B, and the beanbag.

Illustration--1

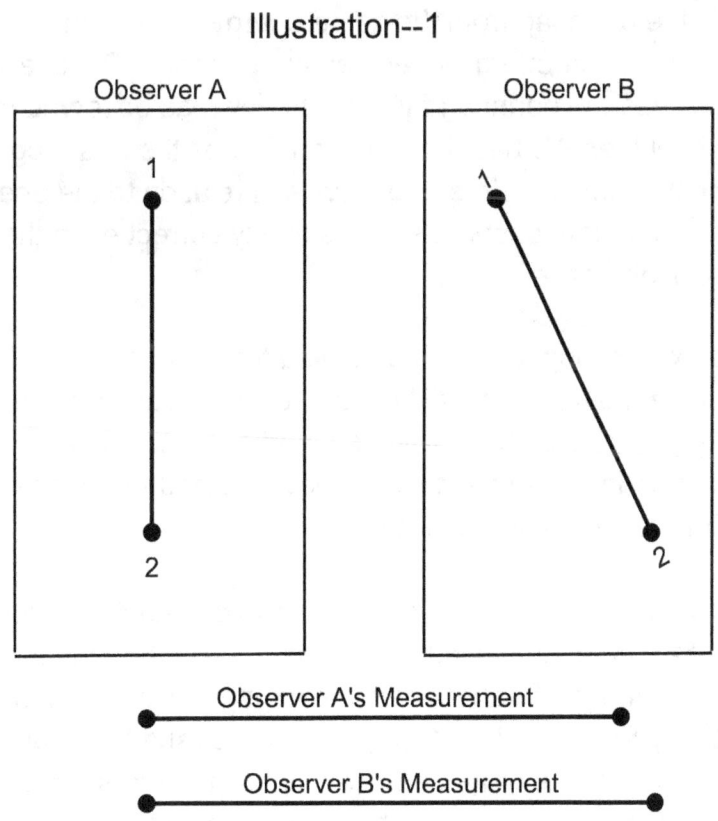

Observer A   Observer B

Observer A's Measurement

Observer B's Measurement

With each incremental increase of speed to the beanbag as it travels in a moving train—relative to earth's inertial frame—the more overall energy the beanbag acquires. For example, if person A drops a beanbag four feet into a pile of sand outside of a non-moving train, the beanbag will displace a certain amount of sand upon impact. The energy acquired by the falling beanbag due to the effects of gravity is transferred to the sand at impact. Now, if the train is moving and person A drops the beanbag the same four feet so it hits the same sandbox that is outside of the train, falling at an angle due to the movement of the train, the overall amount of displacement of sand caused by the falling beanbag is increased. The beanbag literally contained more energy to be transferred to the sand on impact, creating a greater displacement of sand.

This increase of energy is an actual increase of energy that is added within the very structures of the atoms that make up the beanbag. Any change in momentum is a change in the overall energy level of a given mass. This means that the same beanbag in different inertial frames has differing amounts of energy.

How can two different observers, such as observer A and observer B, measure the same falling object to travel two different distances? The difference in distances can be explained on the atomic level. The atoms of the beanbag dropped by person A had the same momentum pattern as the atoms of the moving train and person A on the moving train that dropped the beanbag. The atoms of person B, who was standing on the earth watching the train go by, had the same momentum pattern as the atoms of the earth upon which person B was standing. The atoms of the beanbag, person A, and the train were at a higher energy level than the atoms of person B and the earth. Person A observed the beanbag from the same energy level as the beanbag, while person B observed the beanbag from a lower energy level than the beanbag. Being at the same energy level as the beanbag, person A sees the beanbag fall straight down as observed in Galilean relativity. Being at a lower energy level than the beanbag, person B sees the beanbag fall at an angle towards the ground. From their immediate energy levels in relation to the energy level of the beanbag, the measurements of both observers is absolutely correct even though the beanbag took only one absolute path through space.

When the beanbag was dropped from the moving train, the atoms of the beanbag maintained the momentum pattern of the moving train until impact. Upon impact, each atom of the beanbag experienced a quantum adjustment as they switched inertial frames. In this case, the momentum pattern of each atom of the beanbag lost energy as they adjusted to become synchronized to the new momentum pattern of the atoms of the earth.

In summary, the faster a beanbag (or any mass) moves relative to the earth's inertial frame, the more energy it acquires. The energy level of mass changes when the inertial frame that carries it changes. As a train (or any mode of transportation carrying mass) accelerates to go from one inertial frame to the next, say from 5 mph to 10 mph, its mass (and any mass that appears to be resting on it) literally acquires more energy within its atomic structures. This causes observers of differing energy levels to observe the motion of the same mass differently.

In Stephen Hawking's book, *A Brief History of Time*, Hawking states, "...suppose our Ping-Pong ball on the train bounces straight up and down, hitting the table twice on the same spot one second apart. To someone on the track, the two bounces would seem to take place about forty meters apart, because the train would have traveled that far down the track between the bounces....The positions of events and the distances between them would be different for a person on the train and one on the track and there would be no reason to prefer one person's positions to others" (Hawking, 1988, 17-18.) What Hawking is admitting here is that there is no scientific explanation accounting for the differences in the distances as observed by both observers. Quantum relativity offers an explanation. Each person from their own inertial frame will observe the path of the Ping-Pong ball through space differently because the mass/atoms of each inertial frame has a different momentum pattern and energy level. The person in the train shares the same momentum pattern and energy level as the train and the Ping-Pong ball while the person on the track observes from a different momentum pattern and energy level than the train, the person on the train, and the Ping-Pong ball.

Why do objects that can appear to be at rest within an inertial frame follow the same laws of physics? Whether the train is at rest at the station or traveling at a uniform speed of 100 miles per hour, the Ping-Pong ball falls straight to the ground. What's going on inside the Ping-Pong ball?

Let's put an *atom* on the Ping-Pong table. Because the *atom* seems to be resting on the Ping-Pong table, one might assume that the imaginary horses within our *atom* are at rest, too, that when the train picks up speed, the horses can stand there and enjoy the ride. In other words, when the train goes from zero to 100, one might assume that nothing transpires within the *atom*; it is just a passive passenger along for the ride. But from an atomic model of motion perspective, when our imaginary horses within the *atom* were on the Ping-Pong table at the station, their legs were moving. The momentum pattern of the *atom* is the same as the momentum pattern of the atoms making up the Ping-Pong table, and the Ping-Pong table has the same momentum pattern as the train stationed on the tracks, and the tracks have the same momentum pattern as the earth, to which they are attached. As a force increases the speed of the train, all the atoms that are a part of the train experience changes to their momentum patterns and energy levels. Even the *atom* that appears to be resting on the Ping-Pong table experiences changes in its energy level and momentum pattern as the speed of the train increases. The imaginary horses of the *atom* resting on the Ping-Pong table change their energy level and momentum pattern to match the energy level and momentum pattern of each pair of yoked horses of all the other teams (atoms) connected to the train. All the horses of all the atoms increase their energy level.

In other words, the Ping-Pong ball is not a passive passenger on the train. As the atoms within the train increase in energy, the atoms of the Ping-Pong ball will also increase in energy. Because the momentum pattern of the atoms of the Ping-Pong ball remain synchronized with the momentum pattern of the atoms of the Ping-Pong table and the rest of the train, the Ping-Pong ball will drop straight down if you held it above the table and dropped it. The effects of gravity cause the ball to fall straight down. This phenomenon happens within any inertial frame where all the atoms share the same momentum pattern (the horses are all moving at the same speed and in the same direction).

To help visualize relativity from an atomic model of motion perspective, imagine the following experiment that could be done on the international space station that is free falling around the earth. Imagine Person A floating from one side of the space station to the other side. When he reaches person B, who is stationed in the middle, he lets go of a quarter. Because the individual atoms that make up person A and the quarter share the same momentum pattern and energy level (even after person A lets go of the quarter) person A and the quarter continue to move through space at the same speed and in the same direction. Person B, who is stationed in the middle, watches as the quarter slowly moves away at the same speed that person A is moving away. Person B sees the quarter travel a greater distance from his immediate perspective than person A, who sees the quarter travel with him as if he were still holding it. Even though the quarter travels only one distinct path through space, each observer measures a different length of travel of the quarter from his or her immediate perspective. This is because Person A and the quarter remained at the same energy level, while person B observed from a different energy level.

If Galileo could have measured the energy level of the atoms of any object in his boat cabin while his boat was tied to the dock, he then could have compared that measurement to the energy level of the atoms of the same object when his boat was sailing. The measured difference would validate that as mass goes from one inertial frame to another, the momentum pattern and energy level of the atoms of that mass change to accommodate the new uniform motion. The acceleration and deceleration of mass are only temporary stages between the equilibrium of uniform motion. The acceleration or deceleration of mass is always accompanied by the absorption or emission of energy. This allows each atom to adjust to the new energy level required to maintain its new momentum pattern.

Quantum relativity is the reason the laws of physics are the same for all inertial frames. It also explains why observers in different inertial frames can measure the same moving object travelling different distances through space. These two physical phenomena can be explained by changes in the energy level and momentum pattern of the atoms of the same mass in different inertial frames.

4. Quantum Gravity: The Role of the Atom in Gravity

Gravity is typically explained as an attraction between masses. Quantum gravity explains this attraction as the emission and absorption of energy between masses. The absorbed energy accelerates the momentum of atoms. An acceleration of atoms is an acceleration of mass.

Is an object that is sitting on the earth's surface with no apparent motion really at rest? Newton's observation of distinguishing between objects at rest verses objects in motion has proven fatal for many centuries in understanding how gravity really works. As a result of Newton's mistake, the motion myth paradigm (the myth that mass just mysteriously moves through space without a reasonable atomic explanation as to how or why) has tightly gripped the minds of most, if not all, physicists. And, to this day, physicists cannot explain how gravity operates from an atomic perspective.

When Newton said that an object in motion tends to stay in motion unless acted upon by a force, he was obviously referring to an external force such as a collision or friction. The result of the external force is a change in the momentum of that mass. From a quantum perspective, an external force causes a change in the momentum pattern of each atom, which initiates a simultaneous change in the energy level of that atom. This happens every time an external force acts upon an object in motion. Energy is absorbed or emitted to exactly correlate with the momentum change.

The same effect occurs when energy is absorbed into the nucleus of an atom. A change in the energy level of an atom initiates a change in the momentum pattern of that atom. [Say the horses are spooked and they start running faster (*energy level*) the overall speed of all the yoked horses increases (*momentum pattern*).] The result is the acceleration of that atom in the direction of absorption. When energy is added to the nucleus of an atom, its momentum simultaneously shifts to accommodate the increase in energy. This momentum shift is the acceleration.

Einstein's happiest thought was the realization that inertial acceleration and gravitational acceleration are equivalent. In other words, gravity is the acceleration of mass. He used an example of an elevator being pulled up in empty space at 32 feet per second per second to illustrate this point. Thirty-two feet per second per second means that for every second of acceleration, an object travels an additional 32 feet of the distance of the previous second. At the end of the first second, an object travels 32 feet. For the second second, the object travels 64 feet within that second for a total of 96 feet for the first two seconds. For the third second, the object travels 96 feet within that second for a total of 192 feet for the first three seconds and so on. The elevator being pulled up in empty space would create the same gravitational acceleration that a person experiences standing on the earth or an object falling towards the earth. The reality that gravity is the acceleration of mass creates the basis for quantum gravity.

From an atomic perspective, inertial acceleration is equivalent to gravitational acceleration. Both result in the change of the energy level and momentum pattern for every atom involved. An external force causes inertial acceleration, whereas the absorption of energy causes gravitational acceleration. Whether by an external force or the absorption of energy, the result is the same—the atom or mass accelerates. This is because a change in the momentum pattern (external force) changes the energy level of an atom, and vise versa, a change in the energy level (absorption of energy) changes the momentum pattern of an atom. Either way, the result is acceleration. The simultaneous effect of the momentum pattern changing the energy level or the energy level changing the momentum pattern is like blowing air into a balloon. As the balloon receives more air, the boundaries of the balloon expand. As the balloon loses air, its boundaries contract. As pointed out in the quantum adjustment discussion, an external force can accelerate or decelerate the linear speed of an atom. Energy absorption always accelerates the linear speed of an atom.

Since mass is a collection of bonded atoms, an acceleration of atoms is an acceleration of mass. The constant flow of energy into the atoms of a mass creates gravitational acceleration. The absorbed energy changes the energy level of each atom, which simultaneously changes the momentum pattern of each atom. This accelerates the already moving mass in the direction of absorption.

This is why Newton's misperception was so crucial. If mass is perceived to be at rest when it is sitting on the earth's surface, then the atoms making up that mass are not perceived to be the cause of that mass's motion through space. Newton's misperception assumed that mass is getting a free ride on the earth. This misperception blocks the potential for understanding how gravity really works. By correcting Newton's mistake, we now understand that the atoms that make up any mass are the cause and continuation of that mass's motion through space, (the horses are always moving). What appears to be mass resting on the earth's surface is actually mass in motion wanting to accelerate but experiencing terminal velocity due to the resistance of the earth's surface. If we could shut off the effects of gravity—the energy causing acceleration—the mass would no longer be accelerating into the earth, nor would it be attached to the earth. A good nudge would cause the mass to go off in a different direction than the earth, accentuating the motion it already had.

The effects of absorbed energy causing gravitational acceleration can be seen when an accelerating object decelerates at the moment of impact with a lower energy level surface such as the earth's surface. In the process of reestablishing equilibrium with atoms of a lower energy level, quantum processes transpire. This sets off a chain of events that can be witnessed at the moment of impact as will be explained in the falling penny example.

The Falling Penny

An example of the effects of quantum gravity is dropping a penny to the floor. First of all, as you hold the penny above the floor, the penny is already in motion. It is sharing the same motion as the hand holding it, just as the quarter shared the same motion as the moving astronaut, even after he let it go. When you let go of the penny, the atoms making up the penny absorb energy emanating from the surface of the earth. (The energy emanating from the earth's surface will be addressed later.) The energy level of each atom changes, like the snap of a whip causing each horse to run faster. This simultaneously changes the momentum pattern of each atom, (the whole team is moving at a faster pace). This increases the overall motion of the penny in the direction of absorption 32 feet per second per second until it hits the ground.

When the penny hits the ground, its increased energy level from the absorbed energy that accelerated its momentum interacts with the lower energy level of the ground. At the very moment of impact, the momentum pattern of each atom of the penny changes. Energy is emitted from each atom to accommodate each atom's new momentum pattern. At the initial bounce, the penny is off in a different direction. If not for gravity, the penny would appear to float off in its new direction until a force acted upon it. (To help visualize this, think of an object in the space station that is floating horizontally towards one of its walls. When the object hits the wall, it changes its direction, and then continues in its new direction until a force acts upon it.) As the penny starts to go off in a different direction, it begins to absorb more energy emanating from the earth's surface, changing its direction and accelerating it towards the earth again to repeat the process over and over until the penny finally appears to rest on the earth's surface, with its atoms sharing the same momentum pattern as the atoms making up the earth. During this whole process, the sum of mass-energy in the universe remains unchanged.

When the penny makes contact with the lower energy level surface, there is an immediate change in the energy level of every atom of the penny. The atoms of the surface where the penny landed are also disrupted at impact, causing the absorption and emission of energy according to each atom's equilibrium need. (This is like the sand that was displaced in the sandbox when the beanbag landed in it.) When the penny finally appears to rest on the earth's surface, the momentum pattern and energy level of the atoms of the penny remain synchronized with the momentum pattern and energy level of the atoms of the earth until another disruption occurs such as when somebody picks up the penny.

Why does the penny appear to stay at rest on the earth's surface? Even though it is continuously absorbing energy emanating from the earth's surface, causing it to want to accelerate, its continuous contact with the surface of the earth restricts the acceleration. In other words, it has reached terminal velocity. The energy necessary for acceleration is continuously being absorbed into the penny, but because the momentum pattern of the atoms cannot change due to their terminal velocity, the absorbed energy is emitted at the same rate of absorption.

Drop a penny a few times on the surface of a table. Watch at impact as the penny goes from a higher energy level (the energy acquired to accelerate it towards the surface of the table) to the lower energy level of the surface of the table (the previous energy level of the penny before you picked it up and dropped it). Drop the penny from various heights to compare the amount of energy it acquires from the varying heights. As the penny hits the table and bounces a few times and then vibrates until it eventually comes to rest on the table, you are witnessing the atoms of the penny going from a higher momentum pattern and energy level to the lower momentum pattern and energy level of the atoms of the table.

Gravity is not a mysterious force of attraction between masses, nor is it the effects of warped space-time, but rather, it is energy (forces) acting within matter. It is the effects of absorbed energy on already moving mass. This is why exposing Newton's misperception that objects can be at rest is so important to understanding how gravity really works. Without this step, you cannot understand how gravity is the absorption of energy accelerating mass that already has momentum.

This is why Einstein was unable to unify general relativity (his explanation of gravity) with atomic/quantum physics. Like Newton, he failed to recognize that all objects are always in motion, the first major step for understanding how gravity operates from an atomic perspective. Instead, he brilliantly created space-time as a substitute for atomic/quantum processes, accurately predicting the effects of gravity without addressing the atomic/quantum causes.

Space-time describes distortions in space caused by large masses, such as suns and planets, and the effect this has on moving objects near their surfaces, such as other planets, moons, asteroids, and light. In reality, these distortions are actually higher concentrations of energy emanating from the surfaces of large spherical masses such as the sun or the earth. (This will be explained in more detail later.)

Quantum gravity can now be easily explained as the acceleration of mass caused by the absorption of energy. The absorbed energy changes the energy level of each atom, which simultaneously changes the momentum pattern of each atom, causing acceleration towards the source of the energy being absorbed.

Because the momentum of each proton and neutron is independent of the momentum of all other protons and neutrons (like each set of yoked horses connected together with other yoked horses) the number of protons and neutrons in an atom doesn't change that atom's acceleration rate in comparison to atoms of differing numbers of protons and neutrons (atomic number). All protons and neutrons of a mass are exposed to the same energy source and absorb the same amount of energy, causing the same momentum shift in the direction of absorption. (All the horses increase their speed at the same rate.) This is why Galileo could drop two differing size balls off the same tower and watch them land at the same time. Because the larger ball has more protons and neutrons, its overall energy or inertia will be greater than the ball with fewer protons and neutrons. But the acceleration of both balls will be the same because all protons and neutrons accelerate at the same rate, for all protons and neutrons are exposed to the same energy concentration emanating from the earth. For this reason, all atoms, elements, compounds, and masses accelerate at the same rate. This demystifies the equal attraction law of gravity.

The Visible Effects of Quantum Gravity

The absorbed energy due to the effects of gravity is visible. When I hold a pencil several feet above a hard tile floor and drop it, the absorbed energy responsible for accelerating the atoms of the pencil causes the pencil to bounce a few times when it hits the floor. Energy is emitted from the atoms of the pencil with each bounce until the atoms of the pencil share the same energy level as the atoms of the surface of the floor upon which it now appears to be resting. When I held the pencil above the floor, it is said to have potential energy. Potential energy is nothing more than potential changes in momentum patterns—acceleration—before a terminal velocity is reestablished with the surface of the earth.

When a ball bounces off the ground, it moves away from the surface of the earth. It would continue in this direction uninterrupted if it weren't for energy being absorbed back into the atoms of the ball due to the effects of gravity. After the ball reaches its apex, briefly sharing the same momentum pattern and energy level as the atoms of the earth, it begins to accelerate towards the earth again. With each bounce, more energy is emitted until eventually the ball comes to a resting position on the earth. You can visually see the effects of acquired energy causing the acceleration that precedes each bounce. When the ball appears to be resting on the earth's surface, it is still moving through space. The momentum pattern and energy level of each atom of the ball are still the source and cause of its movement through space. If I shut off the energy emanating from the earth and then threw the ball into the air, it would move away from the earth like a helium balloon escaping from the grasp of a child.

When my car comes to a quick stop, the name tag hanging from my mirror continues to swing back and forth until the energy level of its atoms reaches equilibrium with the new energy level of the atoms of the car of which it is attached to by a string. Here is an interesting point about the swinging nametag. As the nametag's momentum causes it to swing towards the apex of its swing, it slowly comes to a stop. Energy has been emitted so that it temporarily shares the same momentum pattern as the car. At that moment, the effects of gravity accelerate the nametag towards the earth causing it to do a grandfather swing in the opposite direction to its new apex. Gravity then accelerates it again and the process continues back and forth until eventually it evens out and hangs straight down, sharing the same energy level as the car.

As absorbed energy accelerates mass towards the surface of the earth or as emitted energy accompanies the deceleration of mass when it appears to come to an abrupt stop, the effects and results are observable and ultimately measurable.

An example to help visualize quantum gravity is to imagine a wall in space emitting gravitational energy. If you placed a marble several feet away from the wall, it would begin accelerating towards the wall. When it reached the wall, it might bounce a few times and then the marble would appear to rest against the wall. The energy level and momentum pattern of each atom of the marble would be at a terminal velocity against the wall, wanting to accelerate but restricted by the wall. The atoms of the marble would share the same momentum through space as the atoms that make up the wall. In this state, the same applied force would move the marble the same distance in any direction on the wall. (In essence, the wall acts like the surface of the earth.) If you flicked the marble with the same force in any direction, it would go the same distance. If you picked it up and let it go, it would accelerate towards the wall. When it hit the wall, it would bounce a few times, emitting energy until the energy level and momentum pattern of the atoms of the marble matched the energy level and momentum pattern of the atoms of the wall. Then it would again be in a state of terminal velocity against the surface of the wall, appearing to be at rest on the wall.

Tides

Another effect of gravity that is observable but difficult to explain is the cause of tides. Heretofore explained as the gravitational pull of the moon on the earth can now be explained as concentrated gravitational energy emitted from the moon and absorbed by the atoms making up the water. This causes a shift in their momentum patterns, accelerating them in the direction from which the energy was absorbed until a terminal velocity is reached between the energy emitted by the moon in contrast to the energy emitted by the earth. The moon doesn't mysteriously attract the water nor does warped space cause the water to move towards the moon, but rather, it is energy emanating from the large, spherical-shaped moon accelerating the mass of the water towards the direction of absorption.

Satellites

Whether it is our solar system orbiting the center of our galaxy, the earth orbiting our sun, or the moon or satellites orbiting the earth, quantum gravity keeps them in orbit. As the energy cycle keeps the flow of energy moving through the galaxy, spherical bodies absorb this energy and then emit it through their surfaces. The energy emanating from their surfaces accelerates bodies already in motion (within a reasonable proximity of their surface) in an orbital pattern around their spherical shape. It is precisely because these satellites already have motion that energy emanating from the surfaces of large spherical bodies can accelerate their motion into an orbital pattern.

Summary

Gravity is the absorption of energy into the atoms making up mass, initiating and sustaining the acceleration of already moving mass in the direction of the absorption.

Part II

How Spherical Bodies Absorb and Redirect Energy within a Galaxy

How does the earth or any sphere sustain the constant flow of energy emanating from its surface in order to maintain the constant effects of gravity? The earth is already in motion around the sun. The amount of energy it receives from the sun allows it to continuously accelerate towards the sun without accelerating into it. As energy is absorbed into a spherical mass, mass either accelerates or—like the penny experiencing terminal velocity as it rests on the earth's surface—acceleration is restricted and energy is absorbed and emitted at the same rate. In spherical masses, the space necessary for atoms to change their momentum is limited. This means that the constant flow of energy being absorbed by spherical masses accelerate the atoms of the spherical mass until they experience terminal velocity as their momentums are restricted by the other atoms that make up the spherical mass. At terminal velocity, the atoms absorb and emit energy at the same rate. Eventually, the energy emitted by the limited momentum of the atoms that make up the spherical mass has nowhere to escape except through the surface of the spherical mass, and when it reaches the surface, it is then emitted into space. Emitted energy from any mass causes other masses around it to accelerate towards the first mass. This continuous process of mass emitting energy causing other masses to accelerate towards it produces the spherical shapes that we see in the form of stars and planets.

Illustration--2

Energy Emitted from Spherical Surface Area

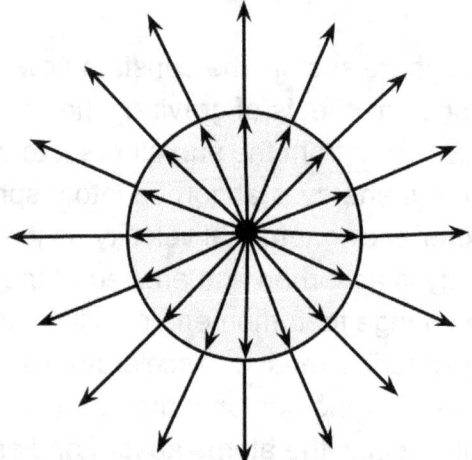

Due to terminal velocities restricting the acceleration of mass within a spherical body, absorbed energy is simultaneously emitted from the atoms of the mass comprising the spherical body until the energy finds its way to the surface of the spherical body and is emitted out into space. This creates concentrated amounts of energy near the surface of a spherical body. As this energy moves further away from the surface, its concentrated amounts are proportionately diluted. Through this process (of energy flowing into and out of the atoms of a spherical body until the energy escapes through surface of that spherical body) spherical bodies provide the source of energy that sustains quantum gravity near their surfaces.

The absorbed energy that keeps the earth accelerating around the sun eventually has nowhere to go or escape except through the surface of the earth, moving out into space. This causes any object on or near the surface of the earth to accelerate towards its surface—gravity. The effects will be strongest closest to the surface and will gradually get weaker as energy spreads out the further it gets away from the spherical surface. Our earth and other planets in our solar system are beneficiaries of large amounts of this energy from the sun. Our sun receives this energy from all sources that absorb and emit such energy like the center of our galaxy, stars in our galaxy, and all the other galaxies in the universe. The center of our galaxy emits enough energy to keep our sun and all the other stars within its system orbiting it. This reciprocal process, now referred to as the energy cycle, keeps energy moving throughout our galaxy and the universe.

In summary, nature has a way of sustaining important processes through cycles. *The energy cycle* keeps a constant flow of energy throughout a galaxy. As part of the energy cycle, energy continually flows into and out of our solar system, keeping the planets orbiting the sun and the moons orbiting the planets. It provides large concentrations of energy near the surfaces of large spherical bodies like our earth, allowing for a constant flow of energy emanating from their surfaces to sustain the perpetual acceleration of masses.

Proof

This book claims that a gravitational energy permeates space. This energy is absorbed and recycled by masses (which are made up of atoms) just as electromagnetic waves are also absorbed and recycled by atoms. This gravitational energy is capable of being absorbed and emitted by the nucleus of an atom and is also capable of being absorbed by electromagnetic waves moving freely through space. The gravity field found around spherical bodies is just larger concentrations of this flowing energy as pointed out in Illustration 2. The gravitational field that causes light to bend near spherical bodies is the result of this energy being absorbed into passing photons. Instead of increasing acceleration like it does when absorbed into the nucleus of an atom, the photon experiences a frequency shift. The speed of light doesn't change, but the shift causes a slight altering of its path. Bent light around spherical bodies is the result of a frequency shift within a photon, (electromagnetic wave).

If the inherent logic of the Atomic Model of Motion is not enough proof for its validity as a working model for the motion of mass through space, then I will offer further proof. The validity of the Atomic Model of Motion hinges upon the starlight displaced by the gravity of the sun. Einstein pointed to the warping of space-time as the cause of this validated phenomenon, (Eddington, 1919). Quantum gravity, as explained in this book, theorizes that the bending of light around large massive objects like the sun is not caused by warped space, but rather, it is caused by large concentrations of gravitational energy emanating from the surfaces of spherical bodies like the sun. As light passes the surface of the sun, each photon absorbs some of this energy, causing a frequency shift. Without affecting its speed, this frequency shift slightly changes its direction towards the source that initiated the shift. The Atomic Model of Motion theorizes that the displaced starlight will be blue shifted from the same starlight that isn't displaced, validating the underlying principle from which the Atomic

Model of Motion is built upon: Energy transfers accompany changes in speed or direction, otherwise, the conservation of mass-energy would be violated.

In Conclusion...

As the quantum model of motion continues to unfold, for this book is only the starting point, the more I appreciate Albert Einstein's Relativity theories. Einstein was one thought away from being able to correct his theories and explain them from an atomic/quantum perspective. Had he realized that mass is always in motion, even when it appears to be at rest, he would have broken the curse of Newton's misperception. He would have been forced to look at the atomic/quantum causes regulating momentum, relativity, and gravity. Although his brilliant theories mathematically predict the results of relativity and gravity, they perpetuate the gap between general relativity and particle physics.

In the end, the purpose of science is not to compose reality to match our perceptions but rather to change our perceptions until they conform to reality—to see things as they are even if it is different from how we want them to be.

Appendix A: Thesis

The crucial years between 1887 and 1905 adversely impacted the advancement of physics. They begin with the Michelson and Morley experiment and end with Einstein's published paper on Special Relativity. During this time, scientists were stumped by the results of the Michelson and Morley experiment. Scientists Fitzgerald and Lorentz both calculated a contraction of the apparatus used in conducting the experiment to explain the results of the experiment. Unfortunately, during these years, scientists missed pursuing the role of the atom in the contraction of the apparatus. In 1905, Einstein's published paper espousing the contraction of time rather than the atom would throw physicists off course even to this day. The Energy Cycle introduces the big picture that Einstein failed to visualize that would have allowed him to explain the role of the atom in momentum, relativity, and gravity. Instead, he substitutes space-time for atomic and quantum processes to mathematically explain relativity and gravity. By failing to explain the role of the atom in his relativity theories, his theories remain incomplete, and the gap between general relativity (gravity) and particle physics still perplexes scientists to this day. The energy cycle bridges the gap between the misrepresentation of space-time and the dice of quantum mechanics.

Appendix B: The Problem with Special Relativity

Quantum relativity would be incomplete without explaining the problem with Einstein's Special Relativity. Einstein's dilemma was similar to the beanbag scenario. Replace the beanbag with a light pulse. Person A, on the moving train, sees a light pulse go four feet. Person B, standing on the stationary earth relative to the moving train, sees the same light pulse travel more than the four feet. (Please note that I understand the above example is not realistic and could not be observed with the natural eye. The scenario is based on the premise that if the distance the light pulse traveled was a lot longer, both observers would measure the same light pulse traveling a different distance.) This is because the light pulse will not only travel the four feet to the same target, but will also travel at an angle proportional to the speed of the train, just like the beanbag. Steven Hawking put it like this:

Since the speed of the light is just the distance it has traveled divided by the time it has taken, different observers would measure different speeds for the light. In relativity, on the other hand, all observers *must* agree on how fast light travels. They still, however, do not agree on the distance the light has traveled, so they must therefore now also disagree over the time it has taken. (The time taken is the distance the light has traveled – which the observers do not agree on – divided by the light's speed – which they do agree on.) In other words, the theory of relativity put an end to the idea of absolute time! (Hawking, 1996, 21-22.)

This scenario only exists if you assume light has the same relativistic qualities as mass, meaning that mass can obtain an apparent state of rest in any inertial frame. Without understanding the role of the atom in quantum relativity, one could easily make this mistake.

To answer the dilemma proposed by Stephen Hawking, I ask the following question: *What is the major difference between a beanbag and a light pulse?* A beanbag can appear to be at rest in any inertial frame. On the other hand, an emitted light pulse is never at rest in any inertial frame. This means that an emitted light pulse follows a different set of rules in how it moves through space than the mass in the form a beanbag. For example, it is common knowledge that the speed of light is unaffected by the speed of the mass emitting it, but the speed of an object (mass) is directly affected by the speed of the vehicle (mass) from which it is released. This major difference is due to the fact that in Galilean and Newtonian physics, objects (mass) can acquire a perceived state of rest within an inertial frame—meaning synchronized momentum patterns—whereas a freely moving light pulse never finds rest within any inertial frame.

Let's quickly diagram the problem.

Illustration--3

Light Clocks

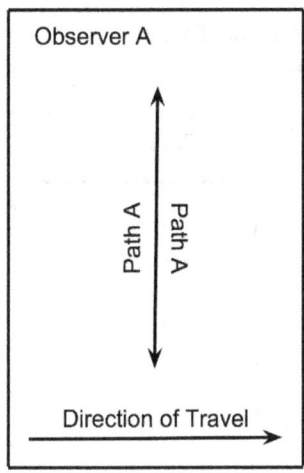 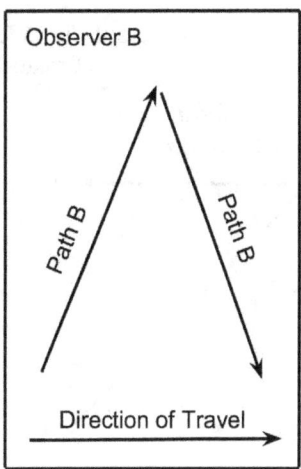

If *observer A* is traveling in a train with a hypothetical light clock, the light will move in an up and down motion in reference to *observer A* (Path A). O*bserver B*, who is stationary to the motion of the train that is carrying *observer A* and the light clock, will observe the path of the light moving at angles in the direction of the uniform motion of *observer A* and the light clock (Path B). The distance of *path A* is different than the distance of *Path B*. Yet, from each person's perspective, they are both correct. This is quite a paradox if the speed of light is constant for all observers and yet travels two different distances. Einstein's idea was to take the different distances of travel and divide them by the same light speed; you end up with two different times for the same event. Mathematically, it makes sense.

Illustration—3 is a classic example used to explain time dilation of Special Relativity in many up-to-date encyclopedias. In illustration—3, if you replaced the light pulse with a bouncing ball, this would be an example of classic Galilean relativity for which I have already provided an explanation on a quantum level. However, a light pulse and a bouncing ball are not the same, so you shouldn't expect the same results. Einstein's mistake was to assume that a light pulse operated by the same rules as mass in Galilean relativity.

Einstein failed to conclude that if the speed of light is independent of the motion of the mass emitting it (Alväger et al. 1964) then its direction or path should also be independent of that motion.

The **speed** and **direction** of emitted light pulses are unaffected by the motion or momentum of the objects emitting them. An object of mass, on the other hand, which can be at the same momentum level as the observed inertial frame within which it is contained, is directly affected by the speed of the object releasing it. Atoms, which make up objects and masses, must be treated differently than freely moving light pulses, which are never at the same momentum level of any

37

observed inertial frame. In other words, freely moving light pulses cannot be regarded in the same manner as masses in respect to Galilean relativity and inertial frames.

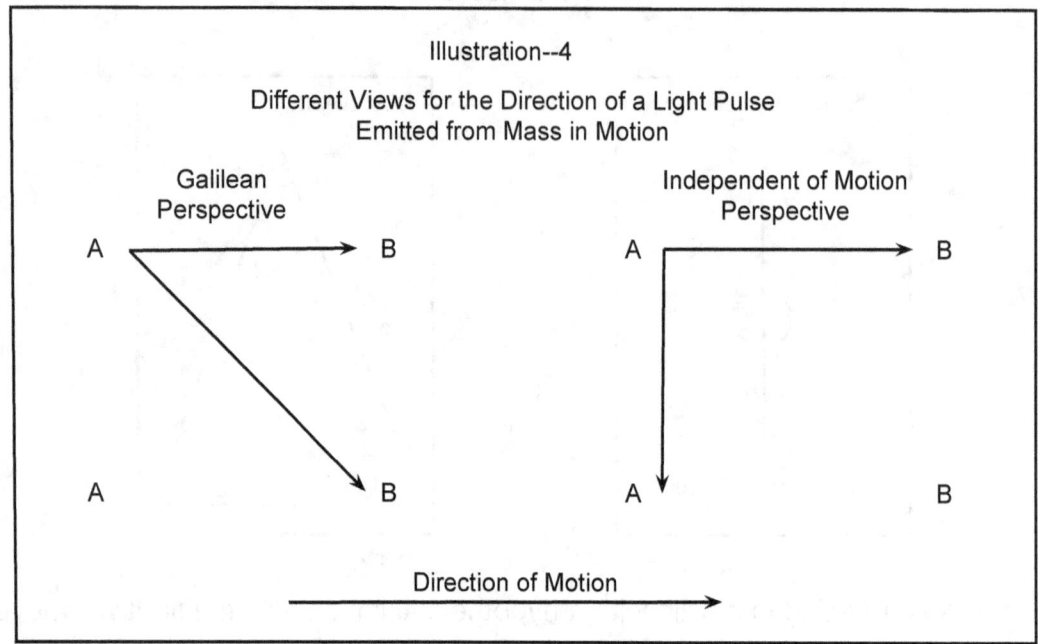

In illustration—4, *A* represents uniformly moving mass in the direction of *B; A* also represents the emission of a light pulse perpendicular to the motion of the mass emitting it. The *Galilean Perspective* demonstrates light's direction or path of travel *dependent* on the motion of the mass emitting it—a continuation of Galilean relativity. The *Independent of Motion Perspective* demonstrates light's direction or path of travel independent of the motion of the mass emitting it.

Just as the **speed** of light is unaffected by the motion of the mass emitting it, the ***path or direction*** it travels after its emission should also be unaffected by the motion of the mass emitting it. This is because light does not have relativistic qualities, whether confined within mass or freely flowing in space. It goes back to comparing the motion of a photon to the motion of an atom. A photon has one fixed speed through space as calculated by Maxwell, whereas an atom's speed through space varies according to the momentum pattern and energy level of that atom. For this reason, when mass is pushed away from a moving body, its new momentum is added or subtracted from its old one like throwing a baseball from a moving vehicle. On the contrary, the speed of light is unaffected by the speed of the object emitting it, such as turning a flashlight on from the same moving vehicle.

I remember someone telling me how Special Relativity was demonstrated in a college class. One student took a piece of chalk and perpetually drew a line going up and down while staying in one place. The second person did the same up and down motions as the first person while walking along the chalkboard from one end to the other end. This supposedly demonstrated how light travels different distances in different inertial frames. Unfortunately, this isn't how light works. The speed of light is not the same for differing inertial frames as postulated by Einstein (who assumed light had relativistic qualities similar to mass), but rather, the speed of light operates independently of all inertial frames. Once light is emitted from the confined energies of inertial mass, its *speed* and *direction* of travel move independently from the momentum of the mass emitting it.

Of all people, why would Einstein make this crucial mistake and then postulate that the speed of light is the same for all inertial frames? He was stuck in a motion myth paradigm, the idea that objects just mysteriously move through space. He failed to theorize the role of the atom in momentum, relativity, and gravity. For Einstein, because objects just mysteriously move through space, Galilean relativity remained a mystery. No one, including Einstein, could scientifically explain why the laws of physics are the same for all inertial frames, (quantum relativity). In his ignorance, Einstein assumed that if the laws of physics are the same for all inertial frames, then the fixed, unchanging speed of light as calculated by Maxwell must also be the same for all inertial frames. It was a terrible miscalculation on Einstein's part that reflected the scourge of the motion myth paradigm.

When Galilean relativity is explained from a quantum perspective using momentum patterns and energy levels, then one can easily see that a photon operates differently than an atom. An atom can obtain an apparent state of rest in differing inertial frames, whereas a photon never obtains an apparent state of rest in any inertial frame. Had Galileo been able to observe the energy levels of atoms, he would have noticed that they change to correspond with differing inertial frames, confirming the role of the atom in momentum, relativity, and gravity.

But alas, in 1905, Einstein's application of time (with the help of the Lorentz transformation) in his paper *On the Electrodynamics of Moving Bodies* delayed the discovery of the role of the atom in momentum, relativity, and gravity.

Appendix C: Einstein Continued…

The following appeared in a recent TIME Book celebrating Einstein's life. In referring to Einstein in the year 1929, "He optimistically told an English Newspaper that his new work proved at last that 'the force which moves electrons in their ellipses about the nuclei of atoms is the same force which moves our earth in its annual course about the sun, and is the same force which brings to us the light and heat which makes life possible upon this planet." The piece goes on to say that "…Einstein's colleagues in the scientific community were not impressed. For one thing, in order to arrive at his new theory, Einstein had violated rules established by his own general theory of relativity. Within a few years even he was obliged to admit that he had failed once more." (Lacayo, 2014, 78) I believe Einstein's failure to visualize the role of the atom in momentum, relativity, and gravity stifled his ability to reconcile his relativity theories with atomic and quantum processes. I believe The Energy Cycle and the Atomic Model of Motion are a continuation of Einstein's desire to unify general relativity and electromagnetism. I believe scientifically validating the role of the atom in momentum, relativity, and gravity is the first step to continue Einstein's quest to discover a unified field theory. This validation would cement the role of the atom in the motion of mass. This includes its perpetual momentum and changes to that momentum, its measurable relativity to other masses, and its acceleration when gravitational energy is absorbed.

I believe the second step is to pinpoint the nature of gravitational energy that drives The Energy Cycle and the Atomic Model of Motion. It is my belief that gravitational energy causes starlight to bend around massive spherical bodies. I believe gravitational energy is exchanged between an electron and the nucleus of an atom. I believe this dialog communicates the necessary energy adjustments needed to reestablish equilibrium when an atom experiences a momentum shift. I believe this dialog is the means of keeping electrons "in their ellipses about the nuclei of atoms," causing the angular momentum of electrons about the nuclei of atoms. I believe gravitational energy permeates our galaxy and the universe and can be found in higher and lower levels of concentration. I believe the effects of gravitational energy are visible such as observing something as simple as a falling penny. Like the effects of wind on trees, even though we don't see gravitational energy, we see the effects it has on other forms of energy. To detect this energy would be the potential means of manipulating gravity.

Appendix D: Time and Waves

Just as Einstein's relativity theories misrepresented the concept of time, I believe quantum physicists need to take a closer look at the concept of waves in the wave-particle theory. "It was the collapse of the wave that most disturbed Einstein. He imagined the wave impinging on the screen like surf on a beach. According to the second viewpoint, a peculiar action-at-a-distance takes place, which prevents the wave from hitting the beach at two or more points at the same time. As a result, the whole wave collapses like a genie into a bottle and beaches itself at one point on the shoreline. Consequently, Einstein favored the first viewpoint." (Wolf, 1981, 121) Time and physical waves are descriptions of quantum processes rather than realities in and of themselves. Whether water waves, sound waves, or waves traveling through a rope, these types of physical waves are manifestations of the interaction of atoms of differing and changing momentum patterns and energy levels. If waves observed on the physical level can be described as manifestations of the interaction of atoms of differing and changing energy levels, then it makes me wonder if we are missing any *hidden variables* (Wolf, 1981, 121) concerning our present understanding of the wave-particle duality theory that would also help us explain how gravity operates from an atomic/quantum perspective. I believe the atomic model of motion offers insights into momentum, relativity, and gravity that have heretofore been bypassed and overlooked.

Here is my last thought concerning the wave-particle theory. The author of *Taking the Quantum Leap* says, "By reducing the size of the hole in our first screen, we cause the interference pattern to spread out even more across our second, receiving screen. The smaller we make the hole, the more the pattern spreads....As we widen the hole again...We observe the interference pattern 'tightening up'...The wider we make the aperture, the tighter the rings become....What we determine depends on the size of the aperture." (Wolf, 1981, 139-140) The resulting patterns that change with the size of the aperture indicate wave-like behavior, but what causes that wave-like behavior? When I hear talk about the size of the aperture the atoms go through, I do not hear talk about the atoms of the mass making the aperture. The closer a moving atom gets to the atoms making the hole, the greater the interaction between the moving atom and the atoms that shape the aperture. I predict that the energy that drives motion and gravity is the same energy that changes the course of an atom going through an aperture, for any change in speed and or direction of a moving atom must be simultaneously accompanied by a change in the energy driving that motion. In summary, I believe energy transfers between atoms in motion going through an aperture with the atoms making up the aperture causes the refractive behavior of the moving atoms as observed on the second screen. The same energy that causes starlight to bend around the sun causes moving atoms to refract when passing by other atoms at a very close proximity.

Bibliography

Hawking, Stephen W. *A Brief History of Time.* New York: Bantam Books. 1988.

Hawking, Stephen W. *A Brief History of Time.* New York: Bantam Books. 1996.

Lacayo, Richard. Time Home Entertainment. *Albert Einstein: The Enduring Legacy of a Modern Genius.* New York. Time Books. 2014.

Wolf, Fred Alan. Taking the Quantum Leap. San Francisco: Harper & Row, Publishers. 1981.